Recovery Reflections
on Life, Faith, Love

Recovery Reflections on Life, Faith, Love

Volume II

Hope Ulch Brown

VANTAGE PRESS
New York

Published by Vantage Press, Inc.
516 West 34th Street, New York, New York 10001

Manufactured in the United States of America
ISBN: 0-533-14371-3

Library of Congress Catalog Card No.: 2002092957

0 9 8 7 6 5 4 3 2 1

In Gratitude

. . . for all the wonderful people I have met and the "little miracles" I have seen on my journey through recovery.

. . . for the "DIVINE INSPIRATION" given to Bill Wilson and Dr. Bob (Sermon on the Mount, The Books of James and Corinthians) in writing the twelve steps of recovery.

. . . for my parents', children's, teachers', family's and friends' continued love, support and encouragement.

Contents

Foreword

People continually complicate their lives. Situations occur, and they are perceived as problems. A life event happens, and in our minds we grow it into a catastrophe. For some reason, accepting life simply as it is remains most people's most difficult task.

Hope has the gift of bringing each issue in life down to its real size. She has the ability to see the most seemingly "big" problems simply. Hope knows how to cut life down into small bites, making it easy to swallow. She sees life through the eyes of Faith in God who controls all things and will gladly provide everything His child needs. Hope knows the secret of practicing gratitude, and of welcoming each person, place, and event as important to the development of the story of one's life.

Years ago, Hope told me that "each day is God's surprise party. All you have to do is show up." The more I have reflected on this statement, the more profound I find it to be. Her writings are filled with such little bits of insight, proving over and over again the truth that the prophets have always known; that the greatest wisdom is most often found in the simplest things.

I invite you to read her musings slowly, savoring each phrase. I'm certain you will find as I have that there is much there beneath the surface, truths that will bring you comfort and strength in your life's journey.

—Bonnie Cameron

Introduction

In 1982, I became the single mother of four children ages six, ten, twelve and fourteen. I was an only child and my mother (who had cancer for nineteen years and FAITH that would move mountains) died. She passed away on a Tuesday, was buried on Friday, and my divorce was final the following Monday. Later that year, my oldest son moved to his dad's until he graduated from high school. With all of these events, I still found myself with a sense of CALM which only comes from FAITH.

I was a stay-at-home mom for fourteen years, although I wasn't idle. Involvement in my children's education led me to co-found Girl's Little League Softball, assist in starting the elementary school Parents' Association, and be elected to the Oxford Board of Education serving as treasurer.

I returned to the work force as a teacher and because my life had been touched by alcoholism and I had sought therapy and joined a twelve-step program for families, I tried to help the students with their substance abuse problems. In the three years I taught, two of our students died. All of this inspired me to return to Oakland University for my master's in counseling. God and I were on the "same page" as far as me getting my degree, however, the outcome of where to use it was HIS PLAN! It was not in a school but in a residential drug treatment center. Two weeks before starting my career as a therapist, I had a double mastectomy for breast cancer. Another miracle!

The six years I spent in residential drug treatment taught me the disease of addiction and co-dependency; the program of recovery and "how it works." I was really sad when it closed, because I had gotten to see the daily miracles of sobriety and how the "promises" of recovery come true. It provided me with some of the most rewarding experiences of my life.

Three months after it closed and I lost my job in 1993, my dad died. He was one of the greatest influences on my life and my BEST FRIEND. I opened my own private practice (which I still have) in Waterford on commercial property he had purchased in 1946 and where I grew up as a child. I had truly come full circle. Again, more TRANSITIONS!

Now, twenty-one years later, I continue my journey in my twelve-step program for families and friends of alcoholics and RECOVERY has been a JOURNEY that I'm eternally GRATEFUL I didn't miss.

This book is a compilation of messages written over the years, that I hope you find inspirational, enjoyable, and beneficial. Thank you for allowing me to share them with you.

—With God's blessings and love,
HOPE

Recovery Reflections on Life, Faith, Love

Dear Lord . . .

You are my COMFORT and my REFUGE,
you are my shelter from the storm;
you are the ONE I can go to,
who is loving and is warm.

You are my FRIEND and also Master,
who will ease my every task;
you're my GUIDE and INSPIRATION,
who will help me if I ask.

You're my solace in confusion,
that is sometimes hard to bear;
you're the ONE whom I can turn to,
and I'll always know you're there.

Lord, Help Me . . .

Lord, help me to change, to let go of "my way"
and help me to live and BE for today.
Help me to know who I am, how I feel;
I don't want to pretend, but be open and real.
Help me to overcome trial and pain,
and look for the good, changing losses to gain.
Help me to reach out and ask for my needs,
and through my mistakes, I'll know what to heed.
I can't change my past and tomorrow's not here;
if I work on today, I'll have nothing to fear.
I'll let go of the guilt, put behind all the sorrow
and not be concerned for the things of tomorrow.
I can't CONTROL others, or how they will be;
please give me the strength just to take care of me.
I'll learn how to love and FEEL and be fine,
by walking with you "one day at a time!"

Dedicated to all those I have known, on this "road to recovery."

The Postcard

I gaze at the fall colors,
of red and green and orange and brown.
My Higher Power sent a postcard,
telling me that He's around.

With each and every color,
I know that He is near.
And as the leaves fall to the ground,
He tells me "have no fear."

Soon these trees will be dormant
and they'll take a winter rest,
and in the spring they'll grow again
and will survive another test.

"I am forever grateful
that He shares this scene with me,
'cause the beauty of all nature is:
it's absolutely free."

It's as if my Higher Power said,
"I'm out here in plain sight;
I've always been here by your side,
whether you were wrong or right."

And as I marvel at those shades,
and the sun and sky above,
I know He's sent this postcard,
with His undying LOVE.

Recovery

My actions are always born in THOUGHT,
And so I must "check out" my mind.
Because if I act irrationally,
Then my thoughts are the UNHEALTHY kind.

So, how do I keep my THINKING
Healthy and on the right track?
I work the STEPS—read "THE BOOK"
And for meetings . . . keep coming back!!!!

Bridge to Shore

My happiness does not depend,
on other human beings,
for if I just depend on them
there's something I'm not seeing.

From the bridge of reason to the shore of FAITH,
is such a wonderful trip.
It really eases up my load,
by loosening my grip.

'Cause happiness is an INSIDE job,
between me and my own maker.
I SEE I must be good to me;
now I'm a "giver" not a taker.

Blind Faith

We often look inside ourselves,
    for the answers by the score,
        when if we'd only trust in God,
           with BLIND FAITH we'd SEE much more.

The Stone

"Let those who are without sin,
    be the first to cast a stone."
    And if we all were HONEST,
    there's not one that would be thrown.

Thoughts

CHRIST SAID . . .
You are to "love one another,"
     in the same way that "I have loved you."
     This isn't simply done through WORDS.
     but is shown in the things that you DO!

Acceptance

And "ACCEPTANCE" is the answer,
    to every problem today.
It has a very calming effect,
    On what I do or say.

If I "ACCEPT" what CAN'T be changed,
    I'm peaceful and serene,
If I DON'T, I bounce right off
    My own balance beam.

I'm out of whack, I'm way off track,
    I'm frustrated as can be,
So I slow down, change my THINKING 'round
    And "ACCEPTANCE" comes to me.

Down to Nothing

Sometimes the Lord, He strips us down,
till there's nothing but us there,
And we could look at this and say,
"It really isn't fair."

And yet, He knows what's best for us,
it's simple but it's true.
The only thing one ever has,
is simply being YOU.

The other things, they all exist,
to enhance the life that's there.
But the greatest GIFT that we can have,
is being WHO WE ARE and share.

If God, He didn't strip us down,
sometimes we'd not discover,
His care, His love, His guiding light,
and how to share it with another.

The sun, the gold, possessions, all,
give life a different view.
But the place where all contentment starts,
is in the heart that's WITHIN YOU.

The Fog

Sometimes when looking at the lake,
I can't see the other side,
Because of the morning mist and fog,
the trees retreat and hide.

But are they really hiding?
No, they're out there in plain view,
and once the fog is lifted,
I can enjoy their beautiful hue.

My life is often like this;
I sometimes can't see what's around,
because of my fog called DENIAL,
what I'm looking for, just can't be found.

And once I can finally admit it,
the picture becomes very clear,
and with TRUST in my Higher Power,
I can ACCEPT life and have no FEAR.

The answers are right there before me;
life's puzzle, it starts taking shape,
because I can now see the pieces,
and CHOICES, I finally make.

I start putting the pieces together,
my life . . . it starts to take form.
Just like the trees coming out of the fog,
I feel strong and healthy and warm.

Just Knock

Whatever will be in my life . . . will be.
And I'll ask the Dear Lord
to take care of me.
I'll do unto others and try to be good,
and things will work out,
the way that they should.
I won't have to worry, be afraid anymore.
Like Christ said, "Just knock . . .
and I'll open the door."

Turn It Over

My father taught me long ago,
worry doesn't solve a thing,
all it does is stir one up,
and ulcers does it bring.
I need to turn to God above,
in the times I have much strife,
knowing it is in HIS HANDS,
and go on with my life.

Alone with God

ALONE we come into the world,
    into a place unknown,
with basic instincts, fear and love,
    and a path we will be shown.

Our family, neighborhood and school,
    we had no hand in choosing;
and these things, they may come and go;
    it's OURSELVES we keep from losing.

Each yesterday becomes our past,
    those days we can't relive;
all we can do is LEARN from them,
    and to ourselves to give.

And as we trudge along this road,
    the one that we call "life,"
through FAITH we're guided from above,
    so we can deal with strife.

There's some of us, we have to learn,
    that alcohol is a drug;
we join AA and learn to PRAY,
    and replace it with a HUG.

How rocky this road ever is,
    depends upon our CHOICE;
when we TRUST GOD and do HIS will,
    it's then we can rejoice.

And when we touch the Dear Lord's hand,
    again we'll be ALONE;
it's with HUMILITY and LOVE,
    we'll approach His heavenly throne.

And at this time, we'll say to Him,
    our mission is completed;
we've done the very best we could,
    and through HIM were not defeated.

The "Gift" of Today

The Dear Lord has given me . . . the gift of today,
and I will embrace it in my own special way.

The "gift" is called "life" . . . to be able to breathe,
to see pretty flowers and the growth of the trees.

To enjoy the warm sunshine, whatever the season,
to be who I am and to act, feel, and reason.

To be good to myself and to have peace and joy,
to be kind to all others, big or small, girl or boy.

To have a good attitude in the things that I do,
to give unto others, and to myself be true.

To look for the positives, and learn from the past,
knowing troubles are small and that pain doesn't last.

This gift is a treasure that was given to me,
and with grace from on high, I was blind, now I see.

What greater a gift than this day that God sends;
I will handle it gently with family or friends.

With FAITH in the Father, for guidance and love,
for this gift I am grateful, it was sent from above.

I Find You

I find you in the sunrise,
In the rain that brings the flowers,
In the tinkling of a bell,
In the minutes and the hours.
In the lightning in the sky,
In the love songs that we sing,
In the memories we share,
In the winter, fall, and spring.
In the sweetness of a smile,
In the book I like to read,
In the moon that shines on high,
In the very tall oak tree.
In the stars that glow above,
In the fire that burns so bright,
In my very plesant dreams,
In the darkness of the night.
In the evening breeze that blows,
In the telephone that rings,
Because you mean the world to me,
I find you in all these things.

You'll Be Safe

The past, it is over;
    The future's not here.
      It's today I'll enjoy;
       The Lord said . . .
          "Have no fear . . .
          I'll take care of your lives.
          There'll be joy and some weeping.
          So take my hand, children,
          You'll be safe in my keeping."

A Gift from God

One day there came the weekend hours,
and I had no plans in mind.
It's like the Dear Lord said to me,
"There's something you will find."

I shared two days with my "best friend"
through a wink and gentle stare;
I found there's places deep within,
that I didn't know were there.

We found a raft, and lily pads,
the trees, the water blue,
the moon, the stars, a gentle touch,
and we got an INSIDE view.

There never is a nothing day;
it's what we CHOOSE to do,
and the best that it can ever be,
is sharing me with you.

And if God sends us days like this,
there's nothing left to fear,
'cause we can simply close our eyes,
and those moments REAPPEAR!

For in our minds, we sailed the world,
shared a song, a prayer, a smile,
and little did we even know,
God had planned it all the while.

Without

Without raindrops, there'd be no showers,
Without the minutes, there'd be no hours,
Without the seeds, there'd be no trees,
Without the nature, no birds or bees.

Without the fields, there'd be no wheat,
Without the beasts, there'd be no meat,
Without the sky, no sun, no moon,
Without the song, there'd be no tune.

Without the dark, we'd not know light,
Without the eyes, we'd have no sight,
Without loved ones, there'd be no home,
Without the woods, no place to roam.

Without the sea, we'd need no sail,
Without the letter, there'd be no mail,
Without our friends, no one to care,
Without our FAITH, we'd have despair.

Without the hurts, we'd not know joy,
Without the girl, there'd be no boy,
And through our lives these things we share,
Because GOD'S LOVE is everywhere.

Letting Go of Fear . . . Means Trusting God

We know the pain we're running from,
yet, FEAR what we're running to.
By putting our FAITH in the Father above,
we'll be guided and shown what to do.

He knows my every want and need;
He loves me 'cause I am His child.
I trust He'll send me what is best
and let go and not be beguiled.

He wants nothing more than my happiness,
gives me peace and serenity.
By walking hand in hand with Him,
I'm FREE now to simply be ME.

I believe the worst is over now;
the past, it can be put behind.
I know not what the future brings;
with BLIND FAITH, the answers I'll find.

I'm not the DIRECTOR of this play;
He has the answers for me,
I'll work the program, follow the STEPS,
and I'll get down on my knees.

Letting go of FEAR means TRUSTING God;
He's with me each step of the way.
I know my future is in HIS hands,
so, I'll PRAY and LIVE day by day.

Transitions

Life is in constant transition;
to it we must learn how to yield,
'Cause we're in the hands of the Father,
just like lilies out in the field.

By living each day for the moment,
enjoying the blessings God sends,
and being forever GRATEFUL,
for our food, family, shelter and friends.

Life is a big bed of roses,
with many a thorn on the way,
knowing we'll grow and we'll blossom
with the pains that come through the day.

For the Dear Lord has PROMISED us blessings;
if only we trust in His love,
we find peace and daily DIRECTION,
the same as He gives to the dove.

So when we "let go" we'll discover,
that "transitions" are simply a fact.
With FAITH in the Father in heaven,
we'll be given the WISDOM to act.

—I wrote this poem on April 12, 1992, not knowing that the
treatment center where I worked would close in October 1992.
God somehow prepares me for events to come and I don't realize
He's done it until later.

Life Is What I Make It

I sit here thinking of life and love
And my Higher Power up above.
We're born, we live, someday we die;
There's things we do, we don't know why.
We make mistakes and through the pain,
We grow and change and aren't the same.
For some of us, the struggle's hard;
We go through battles burned and scarred.
Our feelings sometimes are confused;
They're "up and down," our hearts are bruised.
But errors somehow make us grow,
And we're IMPATIENT 'cause growth is slow.
And then one day we get in touch,
For God had given us so much.
We learn to talk, and feel and share,
And for ourselves we learn to care.
We then take charge of things we do,
And use our FAITH to help us through.
For every day the sun can shine;
If I like MYSELF, the world is mine.

Protect Us, Lord

Watch over and protect them,
the ones we hold so dear,
those who mean the most to us,
who are caring and sincere.

Please guide our actions always,
through each and every day,
so that we know how to live,
and love the Christian way.

Change

We go through life not knowing,
By trial and error we learn;
To CHANGE is not the problem,
The RESISTANCE to change . . . our concern.

For changes, they are required,
It's RESISTANCE that gives all the PAIN;
Acceptance and willingness help us,
So we do not remain the same.

Change is rarely desired . . . often required . . . always inevitable.

Forgiveness

Forgiveness is not an option,
to be given occasionally.
It really is an attitude,
that's consistently part of me.
When I exercise my forgiveness,
NOT condoning the things that were done,
I'm simply not dragging around anymore.
Those people whom I would have shun
If I expect others' forgiveness,
of wrongdoings I have done,
I must be open and WILLING
to forgive everyone . . . not just SOME!

Adapting

Sometimes, when we EXPECT too much,
we find ourselves let down.
When things don't go the way we plan,
our emotions run around.
And so we must be FLEXIBLE,
and able to ADJUST.
Revise our plans, not fall apart;
ADAPTING is a must.

Alpha and Omega

We're the Alpha and Omega,
    the beginning and the end,
        and everything there in between,
            because we're both LOVERS and FRIENDS.

Your Friend

If you ever need someone,
    you know that I am here.
If you ever want to talk,
    you know I'll lend an ear.
If you ever need to cry,
    because you feel let down,
You know that I would cheer you up
    and always be around.
If you're feeling joy inside,
    'cause things are going well,
You know I'd share your happiness,
    so me, make sure to tell!
If you simply need someone,
    to care or hold your hand,
You know that you can count on me,
    that FRIEND . . . I ALREADY AM!

Sun . . . Son

Sun of God shine over me,
    and warm me with your light.
Son of God watch over me,
    and help me with my plight.

Sun of God you heat the soil,
    that living things might grow.
Son of God you warm my heart
    and ease my very soul.

Sun of God you light my path,
    so I will not stumble.
Son of God you guide my steps,
    so I will not tumble.

Sun of God without your rays,
    there would be nothing living.
Son of God without your love,
    there would be no forgiving.

And since the both of you were sent
    from the Father up above,
You feed my body, soul and mind,
    the fulfillment of HIS LOVE.

Choices

Our lives are filled with choices,
about families, jobs and friends.
From the time we get up, 'til we go to bed,
to them there is no end.

Do we want to laugh or cry,
sometimes smile or frown?
Do we solve problems easily,
or find we often compound?

First, we must understand ourselves,
just what is it we can do?
Take a look at the options,
but first to ourselves be true.

We must make choices acceptable;
for us it must be the right action.
We know when they are comfortable,
'cause with them comes self-satisfaction.

Sometimes with problems, we're too involved
and can't see the trees in the forest.
If we don't look at them objectively,
our emotions can destroy us!

There is no one else we CONTROL,
yet, others are a part of our lives.
How much they influence what we do,
is up to us to decide.

The CHOICES we make in the future,
will be better than days gone by.
Because after failures, we have LEARNED,
and have experience on our side.

We'll forever be faced with decisions,
and there's only us to blame,
if life is not going the way we like,
Then it's our CHOICES we'll have to CHANGE!

Growing

I didn't know growing could be so much fun,
 'til I ventured to find "who I am."
Learning behaviors that got in the way,
 now pursuing the things that I can.
Knowing what's harmful and not good for me,
 and changing the "dance" that I did.
Being the person who I really am,
 and feeling the feelings I hid!
For life is an ongoing process;
 success is to really be ME.
By sharing and FEELING the things that I do.
 I'm finally open and FREE.

The Journey

Our relationship is best described,
    as an "itch" that one can't scratch.
Or when one goes to light a fire,
    and cannot find a match.
Or when one goes to "drive" the ball,
    and it falls off the tee.
It's simply called frustration,
    that exists with you and me.

Why do we always do this;
    can't we find another path?
Full of joy and laughter,
    without the pain and wrath.

For going 'round in circles,
    for decades at a time,
Expecting something different,
    there's no reason or no rhyme.

So it's time to end this journey,
    I guess we'll just let go,
So each can now be happy;
    we deserve it, don't you know!

I know I'm worth "recovery,"
    finding me and being true;
I will find MY happiness,
    and I wish the same for you!

I wish you peace . . . contentment,
    for that's what I want for me;
I know that I will find it;
    I'm not blind . . . for now I see!

For the first time in my whole life,
    I'm finally being me;
My insides are feeling better
    'cause I really now am FREE
. . . I'm on a new journey.

Wisdom

W—is WILLINGNESS for open honesty.
I—is INSPIRATION to be all that I can be.
S—is SENSITIVITY to self and others' needs.
D—is DEMONSTRATING change, forgiveness, and good deeds.
O—is OPTIMISM for the blessings FAITH can bring.
M—is MOTIVATION to succeed at anything.

Put them all together, it is very plain to see
that WISDOM is the simple key to peace and
                SERENITY!

Outlook

Sometimes I think and wonder,
    what life is all about?
Sometimes I am so very calm;
    at others I could shout!
I guess that's part of living;
    not everything is smooth.
I must make the most of LIFE;
    if not . . . it's I who'll lose.
I'll have a happy outlook
    and do what needs be done,
I'll grow and overcome the hurts;
    then there's no more need to run,
It's perception that will dictate,
    just how my life will be;
Sometimes there's things that just happen;
    how I REACT is up to me.

Who

It's not where we're going, or where we've been,
    or the things in life we acquire.
It's WHOM we choose to SHARE life with,
    that fulfills needs and desires.

Freedom

Freedom is when one's feelings,
    are not kept inside . . . but out.
Being "open" there's exposure;
    others see and there's no doubt.
They know exactly how you feel,
    and all that you can be.
Whether they accept or not,
    doesn't matter . . . you are FREE!

Goals . . . the Ladder

We must set our goals in life,
a rung up on the ladder.
If we cannot climb real fast,
it really doesn't matter.
Just so we're progressing up,
and not falling behind.
We learn to take each forward step,
just ONE rung at a time.

The Singles' Creed

"WE WILL SURVIVE."

There may be things that go awry,
There may be times that we might cry;
Some things have happened, we don't know why,
      But, we will SURVIVE.

There are days when all seems wrong,
It's at these times, we must be strong;
How will we ever get along?
      We KNOW we will SURVIVE.

We will ASK the LORD above,
To DIRECT us with His love;
Our feelings, we'll learn not to shove,
      With HIM, we will SURVIVE.

We can make new happy days,
If we learn to CHANGE our ways;
It's just a temporary phase;
      We know we will SURVIVE.

We know that this is not the end,
Because we have made such good friends;
There is no need now, to pretend;
      They'll help us to SURVIVE.

It's when our lives are changed around,
And we have both feet on the ground;
Happiness then can be found,
      Because we have SURVIVED.

With the CHAOS . . . we're all through,
Our lives, we can begin anew;
And to OURSELVES, we will be true;
      We're sure we will SURVIVE.

On this path . . . we find success,
For us, we now have passed the test;
There is a future of HAPPINESS!
　　　It's GREAT to have SURVIVED!
　　　(Our life is now ALIVE!)

Power

The POWER of the fellowship,
   is greater than me;
      That's why the twelve steps,
         all begin with "WE."

Focus

I know not where my journey leads;
    God is with me all the way,
        and I know I will be healthy,
           if I just FOCUS on today.

My Life

Where am I at in my life today?
    Sometimes, I guess it's hard to say.
My parents are no longer here;
    My children are both far and near.
My past has taught me many things;
    I know not what the future brings.
Through my recovery, I found me;
    My world is what my mind perceives.
I see the good, have gratitude;
    I've learned to CHOOSE my attitude,
By NOT REACTING, I think things through,
    and do the things I need to do.
Emotions are a part of me,
    To acknowledge them, has set me free.
By allowing others to BE and DO,
    I have "to my own self been true."
My FAITH, it is the beacon light,
    Which guides my path and gives me sight.
My source of COURAGE, strength to bear,
    Comes from my GOD, who hears my prayer.
This miracle no one else could give,
    This LIFE God gave for me to LIVE.

His Choice

Once there was a little boy,
with "twinkling" eyes so brown.
He had no other siblings;
all adults he was around.

Then he moved one day to Grandma's
'cause his mom was never there.
The adults they did their own thing;
they were always off somewhere.

So the little boy felt lonely,
and he thought, "What shall I do?"
He couldn't count on others,
so he copped an attitude.

His school work took a tumble;
he was angry, scared and sad,
Mom was not there for him,
and God knows where was his dad.

He was always there for others;
double messages he got.
No one ever listened to him,
so he drank and drugged a lot.

For there he found an outlet;
it would cover up the pain,
'cause the world that he was raised in,
didn't seem to him too sane!

But underneath the chaos,
there was a "heart of gold,"
a caring, gentle little boy,
who just needed to be told:

"Son, you are a sweetheart,
all your dreams, they can come true;
study hard, shoot for the best,
you are SPECIAL being you!

"You have so many talents,
you are fun and kind and smart;
start today, be WHO YOU ARE,
express your feelings from your heart."

So the young man kicked the habit;
he found GOD and love and friends,
and WALKED the WALK, and loved himself;
and it's HIS CHOICE how the story ends.

More Thoughts

REMEMBER THE PAIN
Thank you for the happiness,
      but I'll not forget the pain,
         because if I remember it,
            I won't make the mistakes again!

I DON'T KNOW
I don't know where I'm going, yet,
      but I sure know where I've been,
         and a new way has got to be better,
            so I'm not going that way again.

SOMETIMES
Sometimes there has to be an END,
      for there to be a BEGINNING.
         What often appears to be a LOSS,
            can actually be a WINNING.

TIME
The only thing I've ever asked
      is that you spend some TIME with me.
         This doesn't seem so hard to do,
            since it's ABSOLUTELY FREE!

We Write the Script

We start life as a "blank page,"
many sentences we will write.
Some of them will be okay,
and some right . . . out of sight.

The secret is . . . that we CORRECT,
the mistakes that we have made.
It is through this simple process,
we improve our DAILY grade.

It is only after ERRORS,
and corrections by the score,
that we have a "FINAL COPY,"
of a life that means much more.

The Recovering Child Within

There is a little child in me,
I've kept her deep inside;
because of hurts that she's been through,
it seemed more safe to hide.

She's now grown up and changing life;
she's in recovery;
by letting go of negatives,
she's learning to be FREE.

Her Higher Power is to her,
a beacon in the night.
God guides her down the path of life,
through pains and hurts and frights.

That little child can LEAN on Him,
for strength, and have no doubt,
her God will help her through her trials,
and things will all work out.

She's learning to RISK "letting go,"
trust her Higher Power will be there,
taking care of what she can't control,
knowing problems they will share.

And what is it she learns each day:
"to thine own self be true."
She always understood this phrase;
now she's learning HOW to do.

She's taking "one day at a time,"
now living her life straight.
And through the program . . . step by step,
life's now becoming bright.

To Someone Special

I wanted to write you a message,
and I know what I wanted to say,
that you can find peace and contentment,
if only you'd learn how to pray.

You'll resist when I mention this to you,
'cause you're filled with such anger today.
It's hard for me, watching you do this,
when I know there's an easier way.

The past, it is over and done with,
yet you carry the scars everywhere.
The pain that it caused deep within you,
made you give up and not even care.

If only you knew how important,
it is to let go and to grieve,
that tears are not weakness . . . they're healing,
and then, in yourself you'll believe.

That the love and the heart that's within you,
are worth more than money or gold.
And life, it is really worth living;
it can't be bought, traded or sold.

And once you RISK letting the pain out,
you can finally feel happy inside.
There'll be no more need to be angry,
and distance or run, drink or hide.

The only thing I can do for you,
is care and then PRAY from afar,
that God gives you the courage to feel things
and realize just how "SPECIAL" you are!

The Fire of Love

As I gaze upon the sunset,
    against the sky of blue,
A brilliant orange, round ball of fire,
    it is a gorgeous view.
An awesome splendor to behold,
    yet, I feel contentment too.
Something like the fire that burns,
    in my heart, with love for you.

Love

What in life is important?
It's the thing that we call "love."
It's what is so essential,
Like the sun and moon above.

It's found in daily growing,
In the things we share together,
Willingness to communicate,
That makes life so much better.

It's all the little kindnesses,
The tenderness expressed.
The feeling of contentment,
That brings total happiness.

It's the little imperfections,
That we can overlook.
It's gentle warmth shown in a smile,
The compassion in a look.

It's the caring thoughtful feelings,
That are shown in many ways,
Acceptance, giving, trust, support,
"Best friends" throughout the days.

It's the things that we experience,
Somehow so naturally,
Ability to be ourselves,
That somehow sets us free.

It truly is fulfilling,
This love that we can share;
We know it's irreplaceable,
'cause we can't buy it anywhere.

You're . . .

You're the malt that's in the milkshake,
    You're the peaches in the cream.
You're the petals in the flower,
    You're the twinkle in a gleam.
You're the lyrics in a love song,
    You're the salt that's in the sea,
You're the flame that's in the candle,
    You're the sugar in the tea.
You're the honey in the beehive,
    You're the syrup from the tree,
You're the gold among my treasures,
    'cause you're JOY and LOVE to me!

My Dad

My Dad, he meant the world to me,
    no other of his kind.
You could search the whole world over,
    not another would you find.
He was the baby boy of ten;
    his Dad died when he was three.
His Mom, on whom the burden fell,
    would bounce him on her knee.
She left him at an early age;
    he was eighteen when she died.
He had to face the world alone,
    and he faced his life with pride.
For as a child, he read a lot
    and skipped two grades in school,
And from his large farm family,
    he learned the "golden rule."
One year of college he pursued,
    and because of the Depression,
He had to make his way in life,
    so, he began a new profession.
Because he was so likable,
    the "salesman" he became,
And at my mother's flower shop,
    he sold her his own name!
They married and they settled down,
    in a business run together.
They were a team from whence they met,
    and "love" guided them forever.
And as the only child they had,
    they shared their love with me.
My Dad, he dearly loved my Mom
    and together, we were three.
Though Dad was strong and smart and fun,
    he was very sentimental.
His "wisdom" was what guided me,
    and his love was very gentle.

He had so many attributes,
    his HUMOR most of all.
We'd sit and talk and laugh and sing;
    to me, he "just stood tall."
I learned so very much from him,
    'bout life . . . we'd philosophize.
You knew the being that he was;
    you could see it in his eyes.
And as a GRANDPA . . . he was GREAT!!!
    and is surely missed today.
No greater man, to us has lived,
    is what we all would say.
The Dear Lord took my Mother first,
    my Dad in '93.
But the WISDOM, FAITH, and LOVE they left,
    was to us, their LEGACY!